Flying Pictures

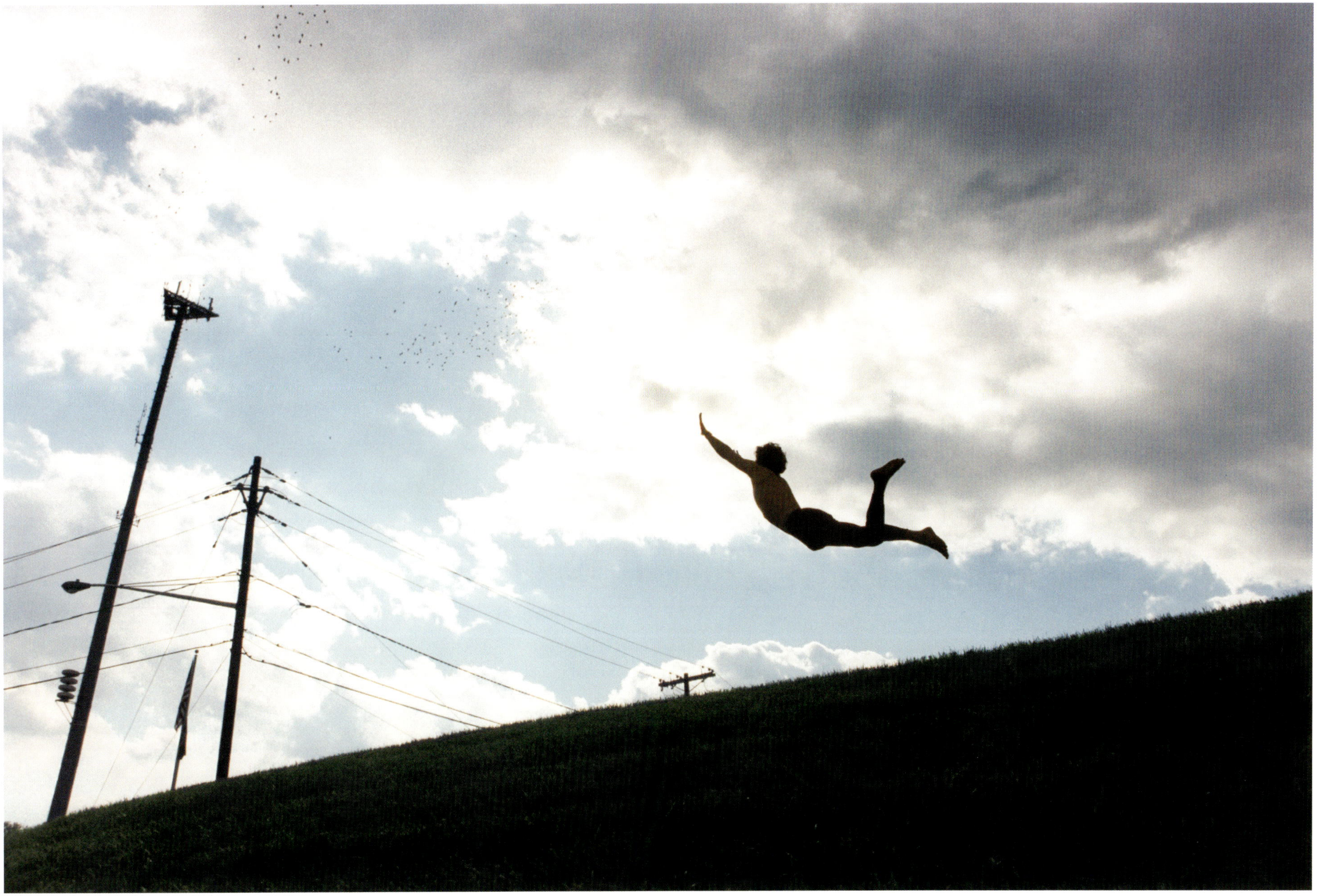

Flying Pictures

Daniel Gordon

Foreword by Gregory Crewdson

pH powerHouse Books Brooklyn, NY

what can his stories of silent flights tell me
of solid things that vanish in a magic smoke?

—Don Geiger

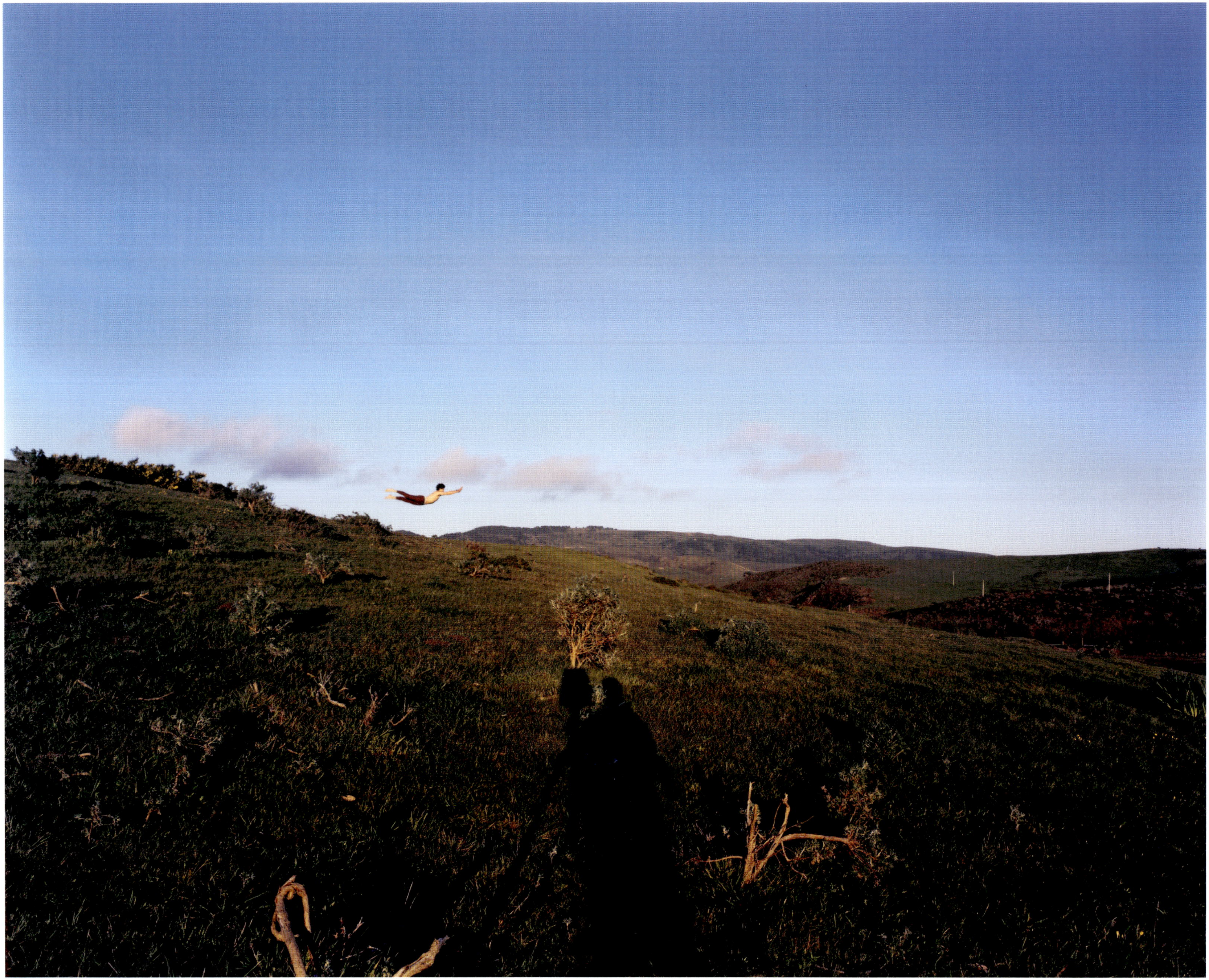

A Note on the Pictures

As a young artist, Daniel Gordon set out to learn how to fly. He approached this endeavor with a great degree of ambition, athleticism, and determination. This dedication went hand in hand with a sense of wonder and possibility. Playing with chance while ardently pursuing his task, he merged unwavering obsession with irrationality; the very foundations of an artist's practice.

The medium of photography has always had a direct relationship with truth and the representation of facts. These pictures present an intersection between Daniel's compulsive desire to fly and photography's capacity to capture the moment.

Beautiful in their simplicity, they are fundamentally an observation on the relationship between figure and space; and a romantic view of a luminous, pastoral landscape.

On a fundamental level, the art of photography is about the miracle, and how to make the ordinary extraordinary. These pictures do that in it's purest form. They capture suspended moments, perfectly situated between transcendence and doom.

Flying is a young man's game and this work was made over a short and fleeting period of time. It is filled with a sense of innocence and hope that is, in its very essence, impossible to hold onto.

Gregory Crewdson
February 2009

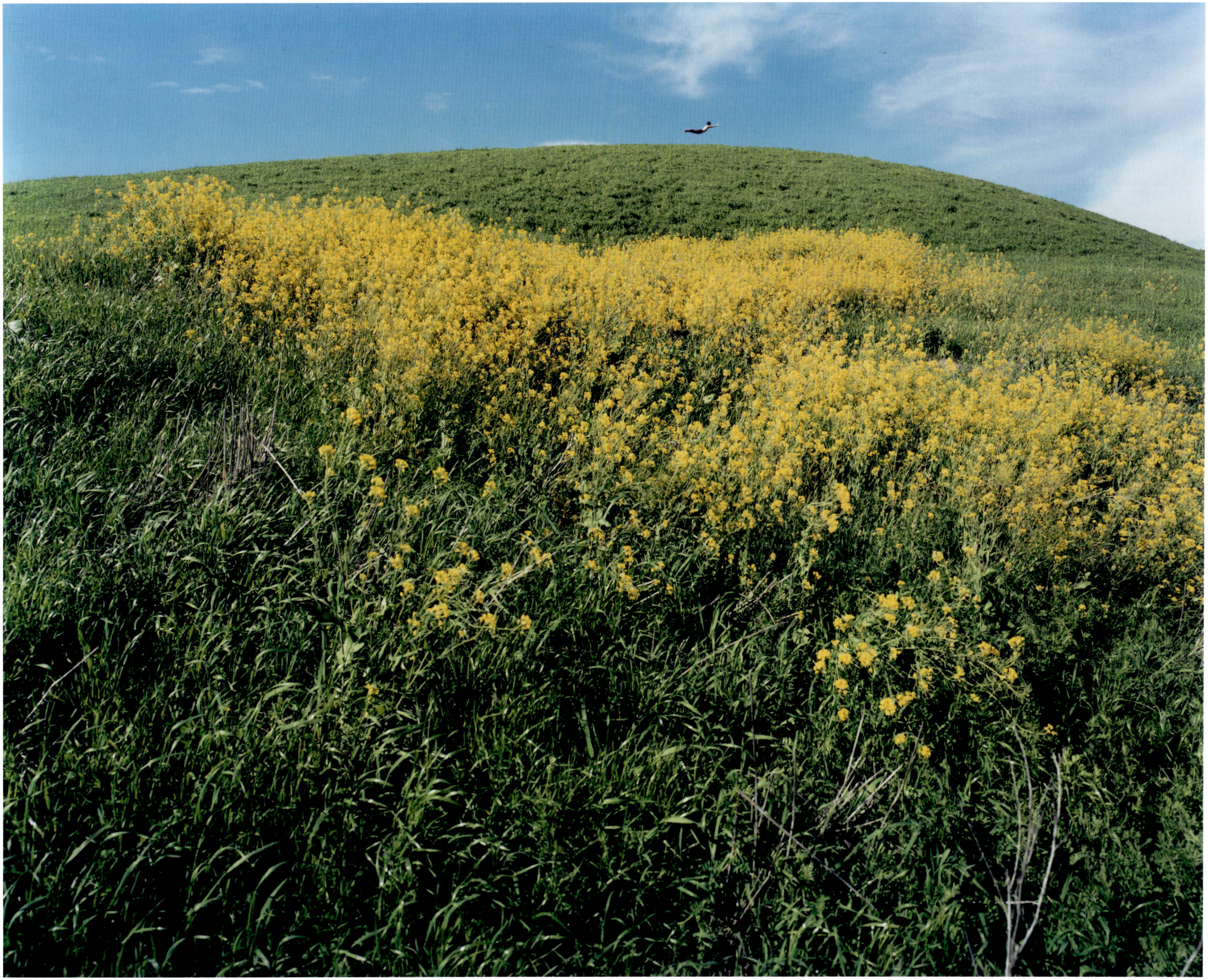

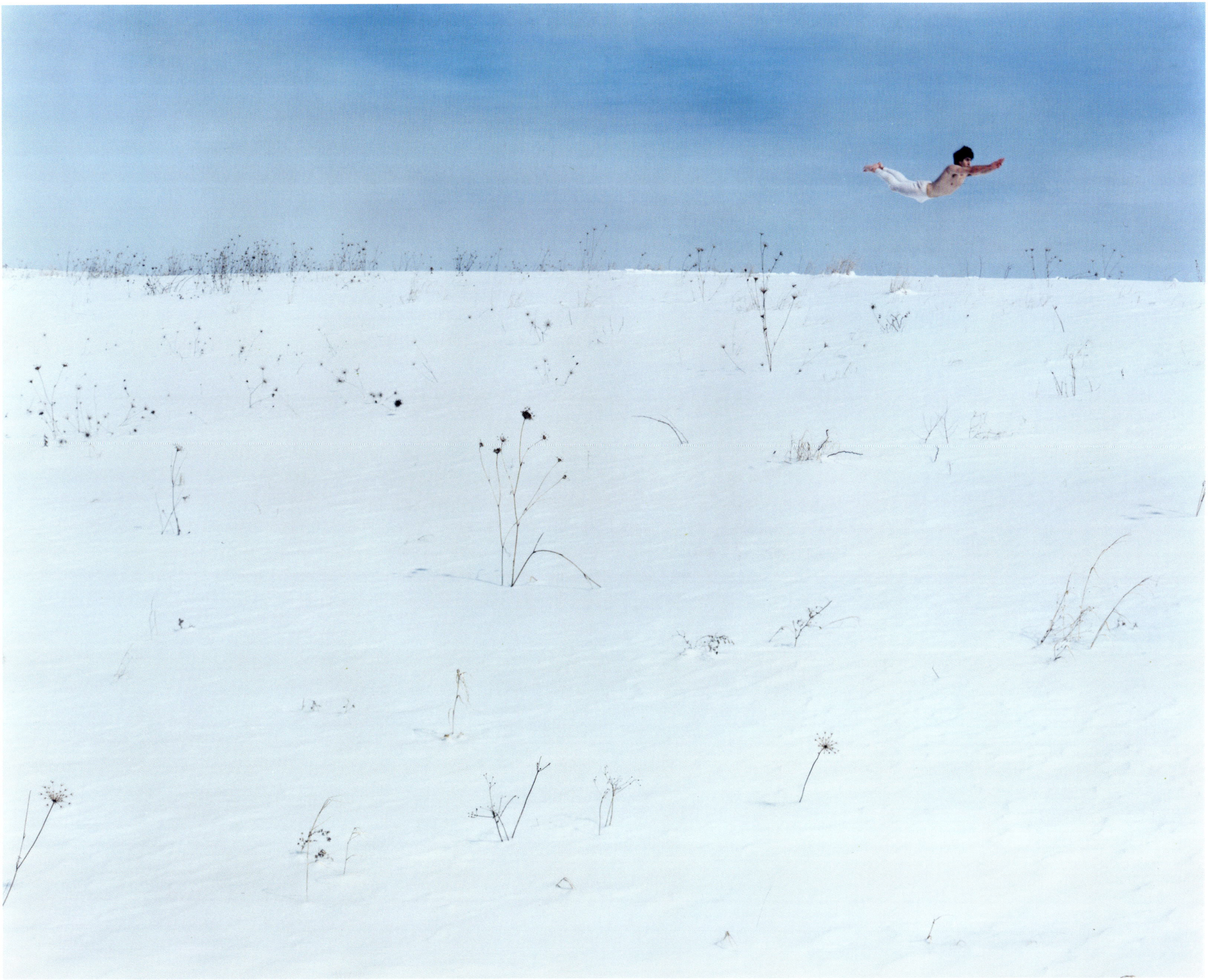

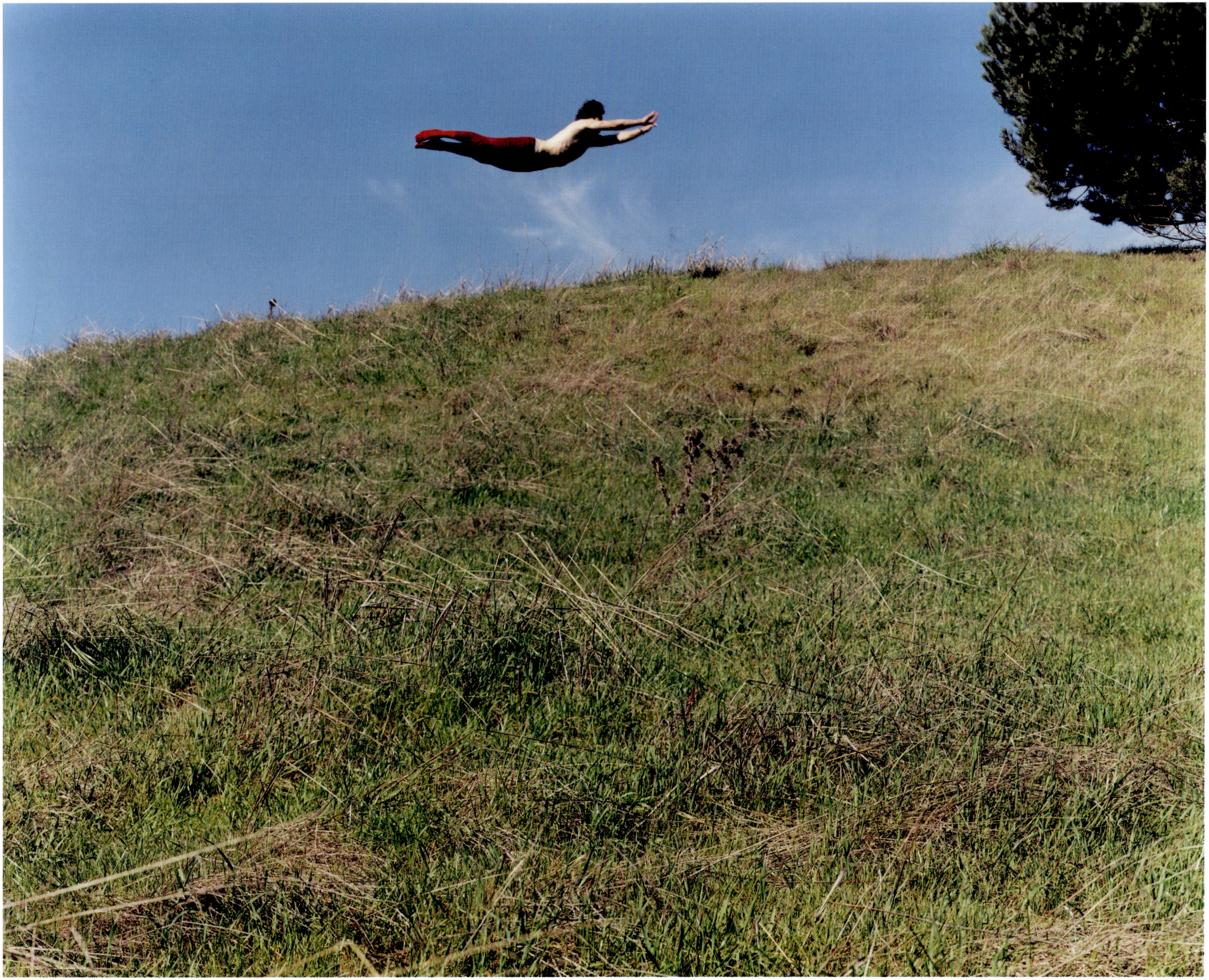

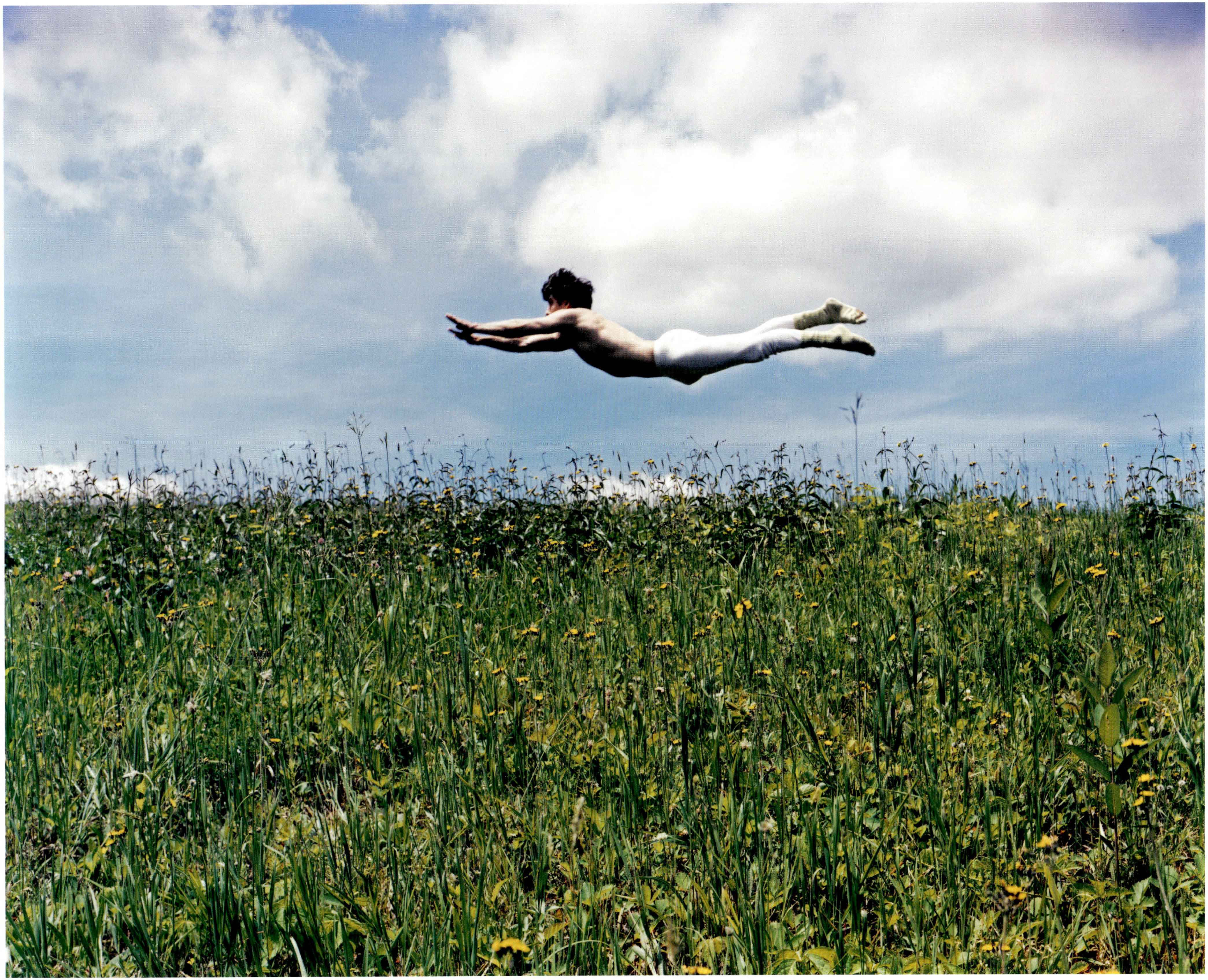

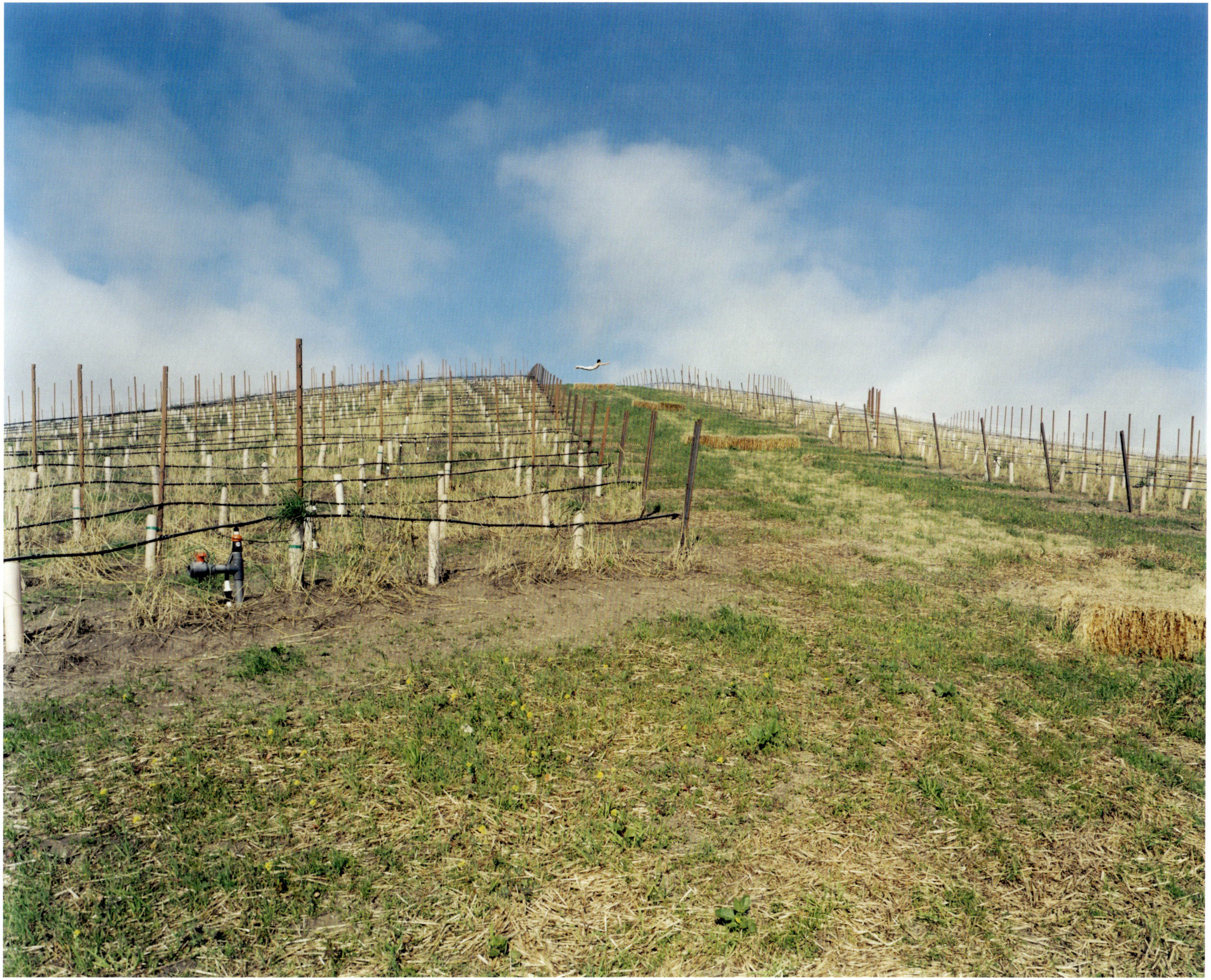

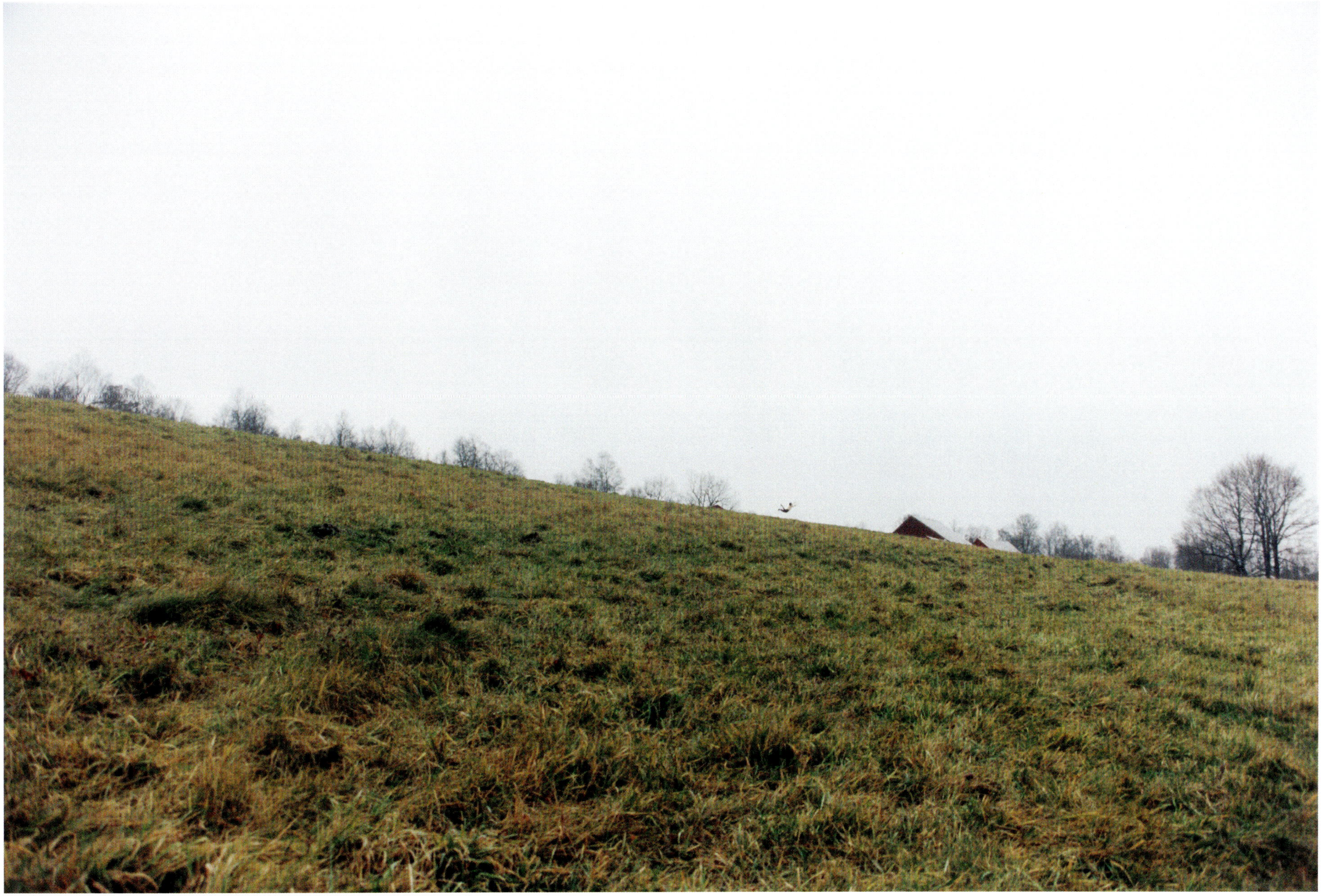

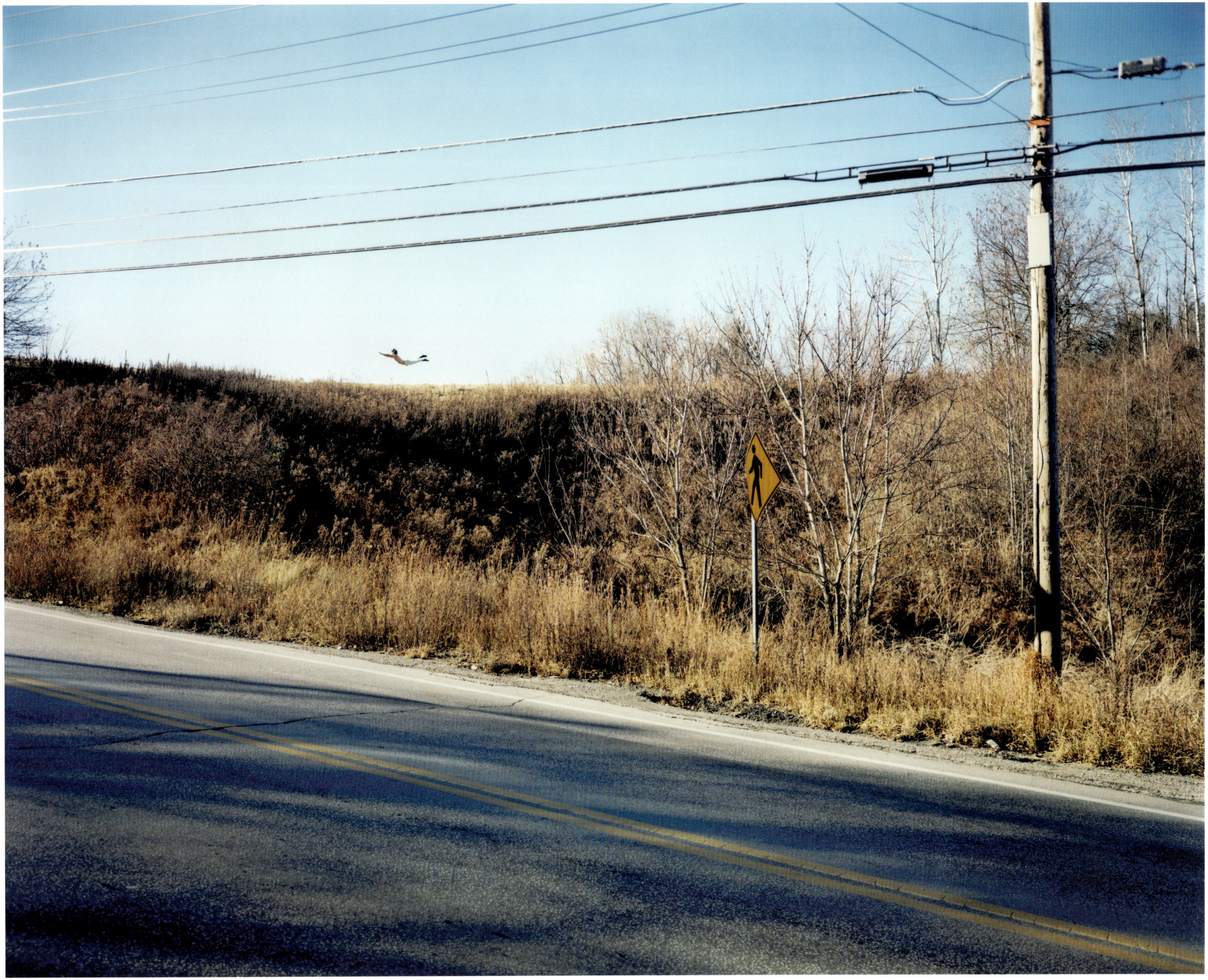

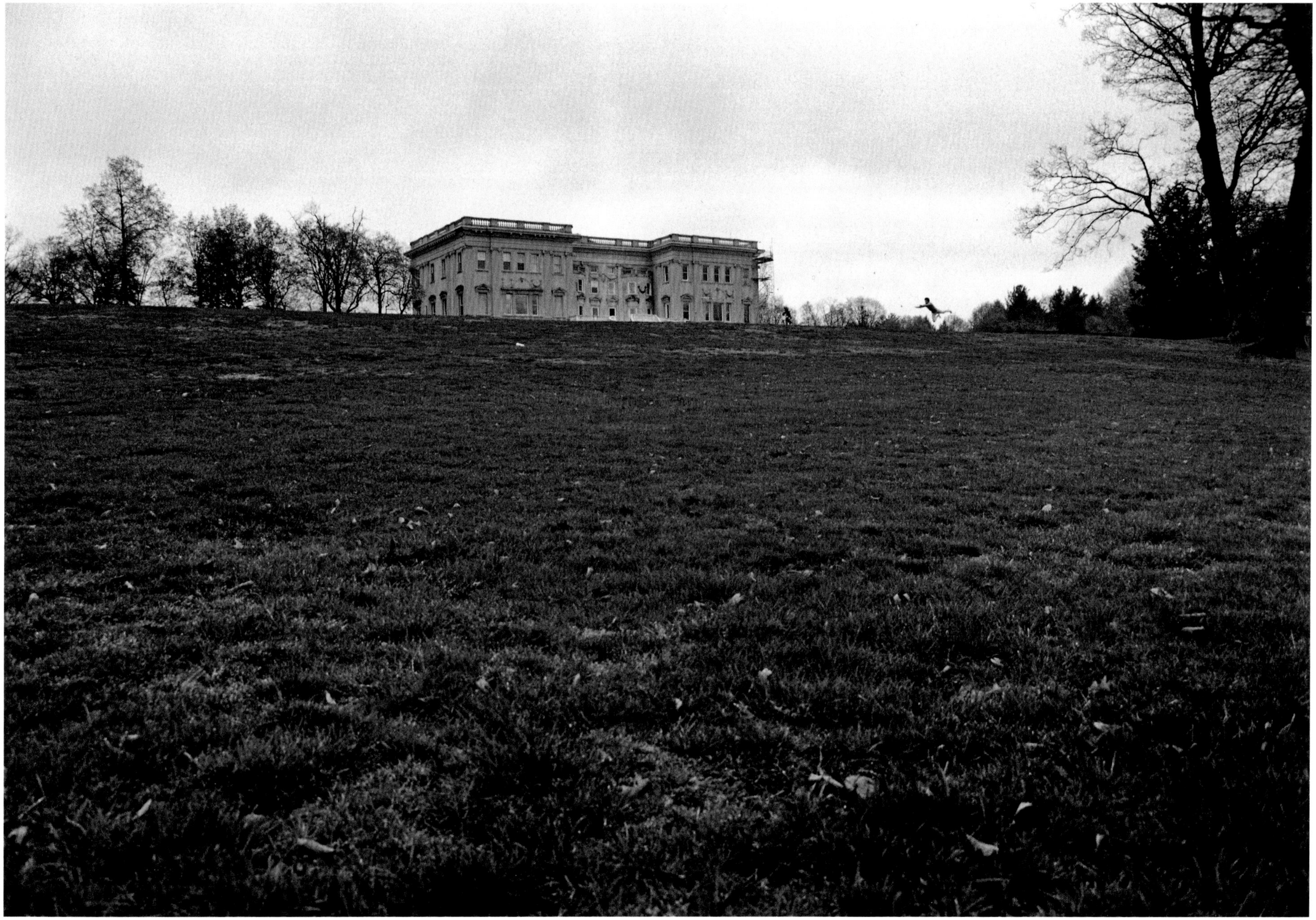

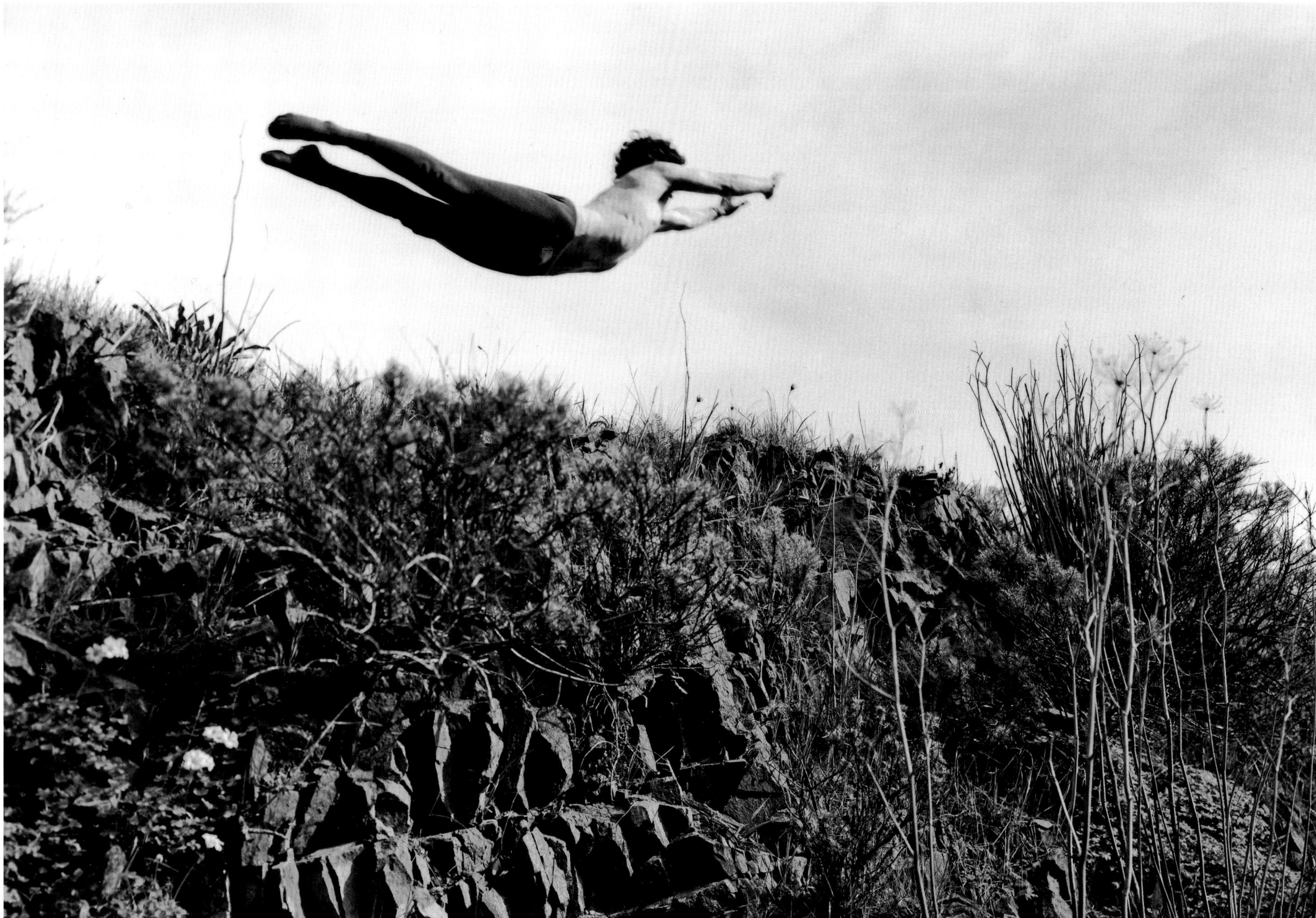

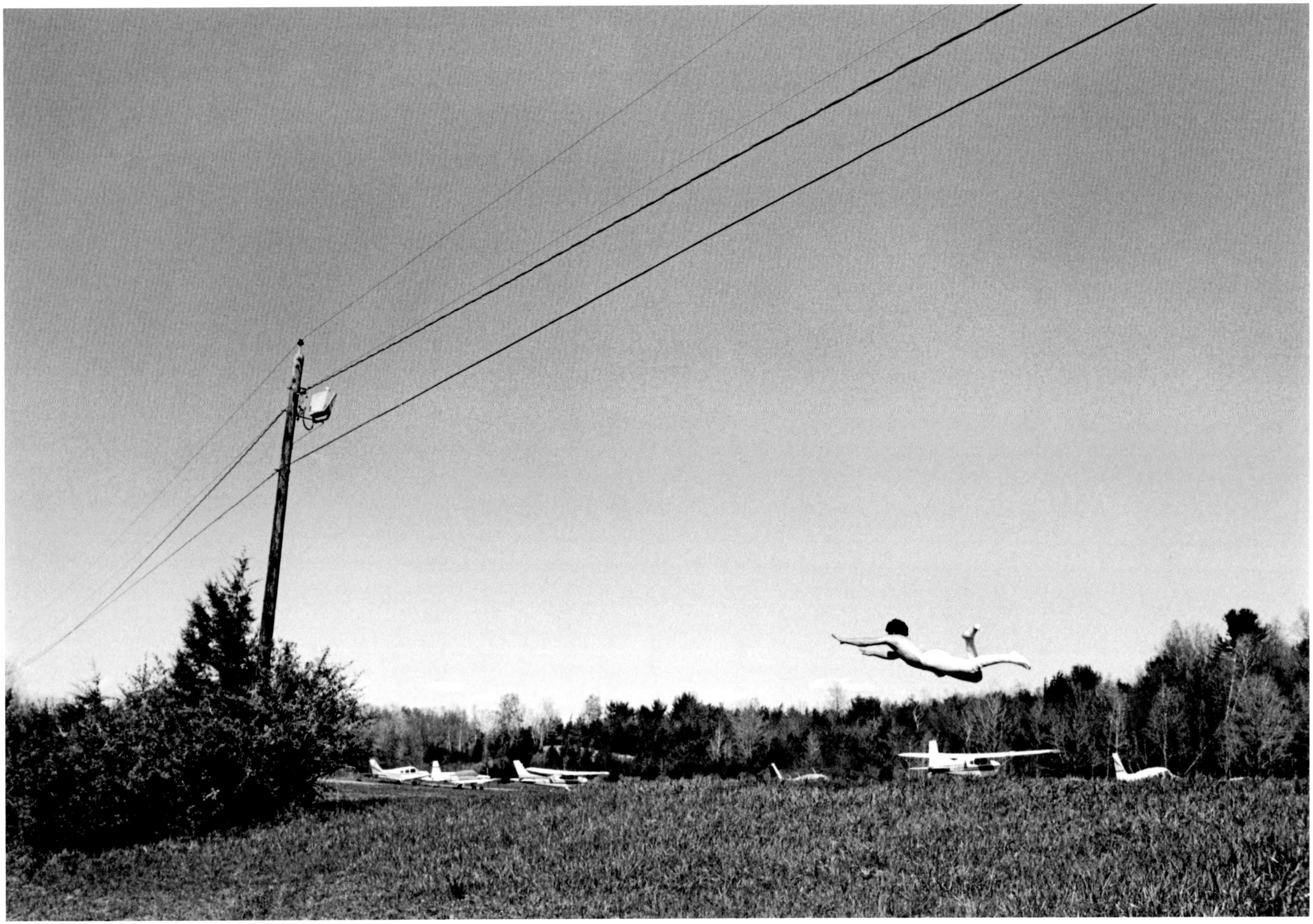

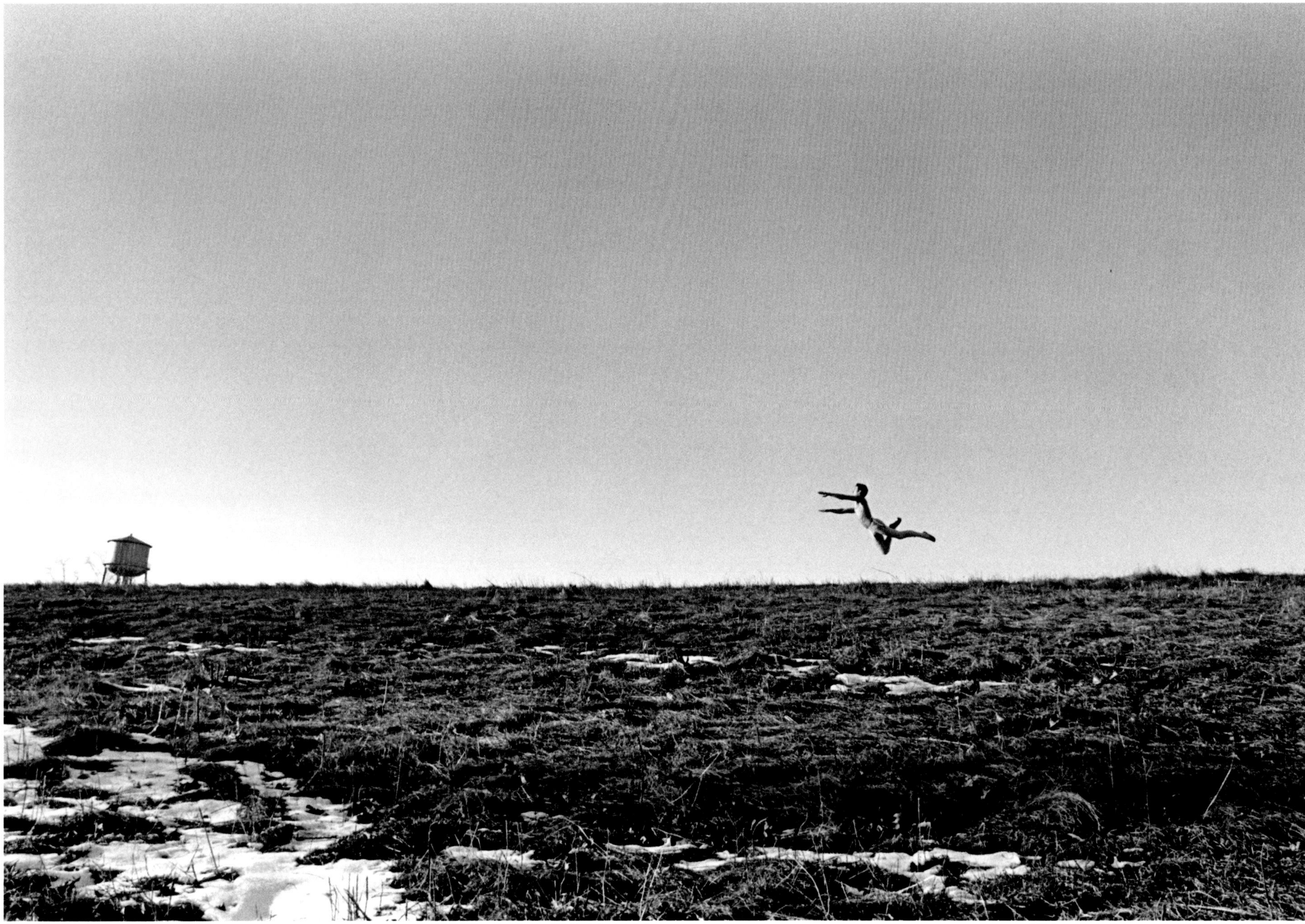

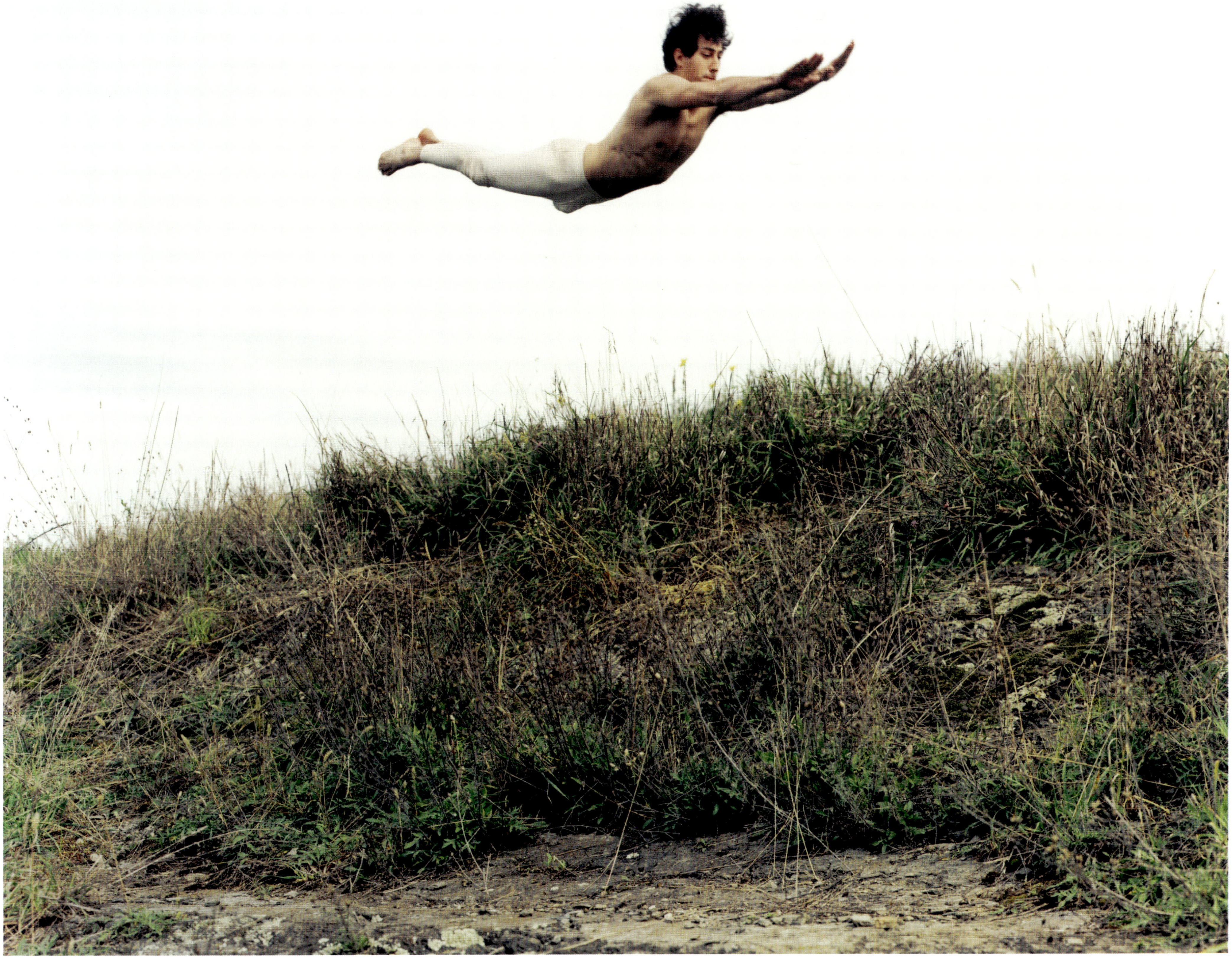

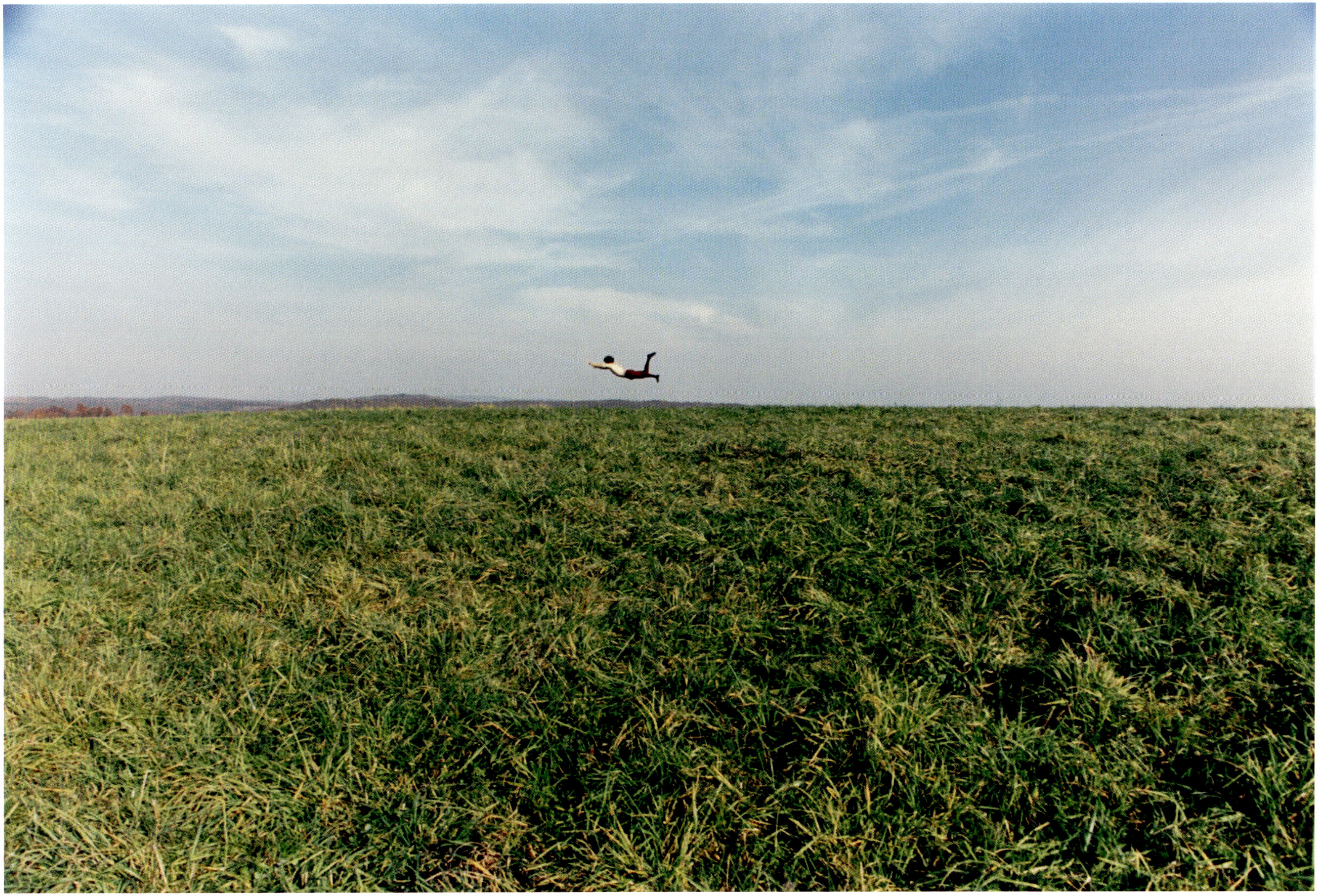

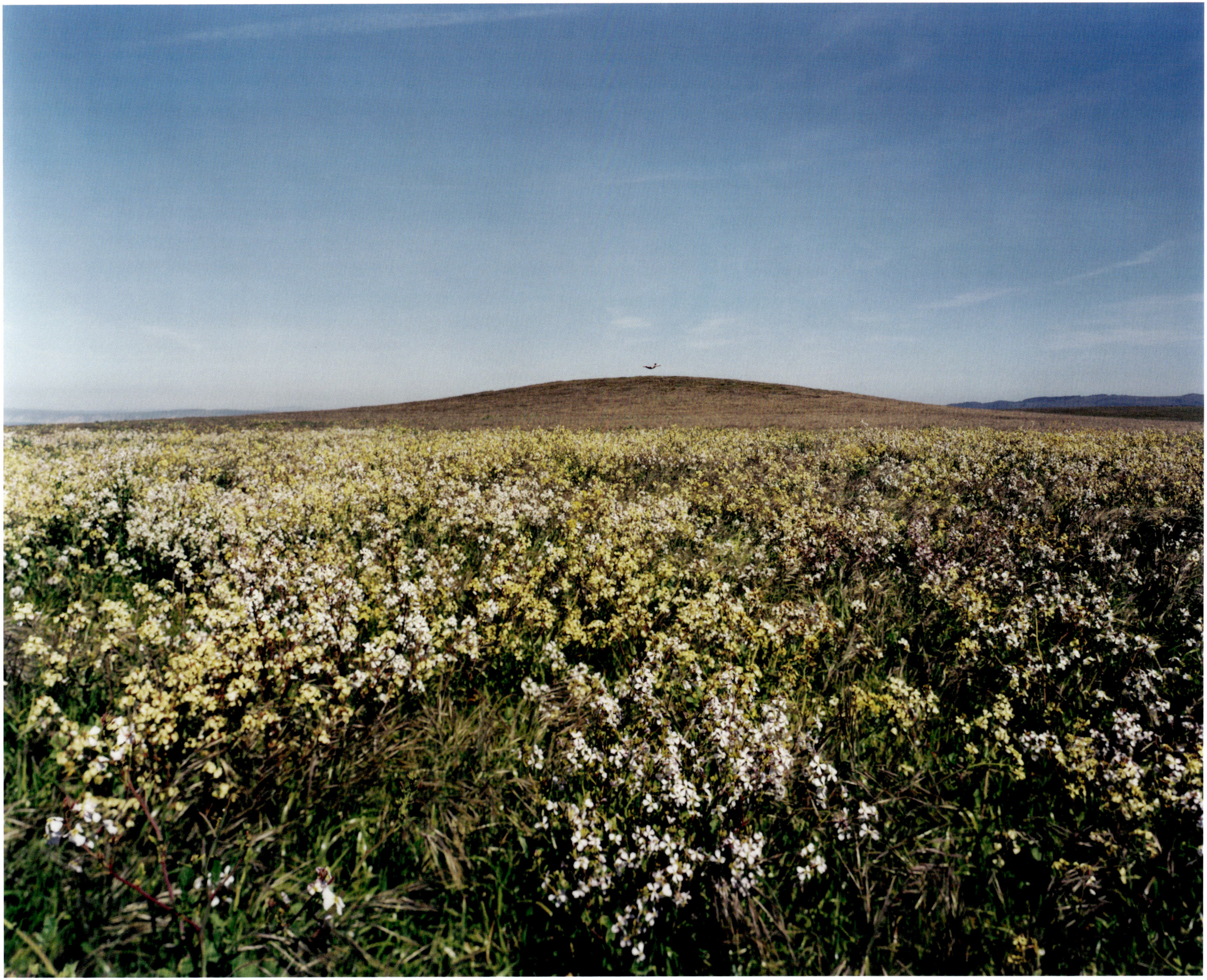

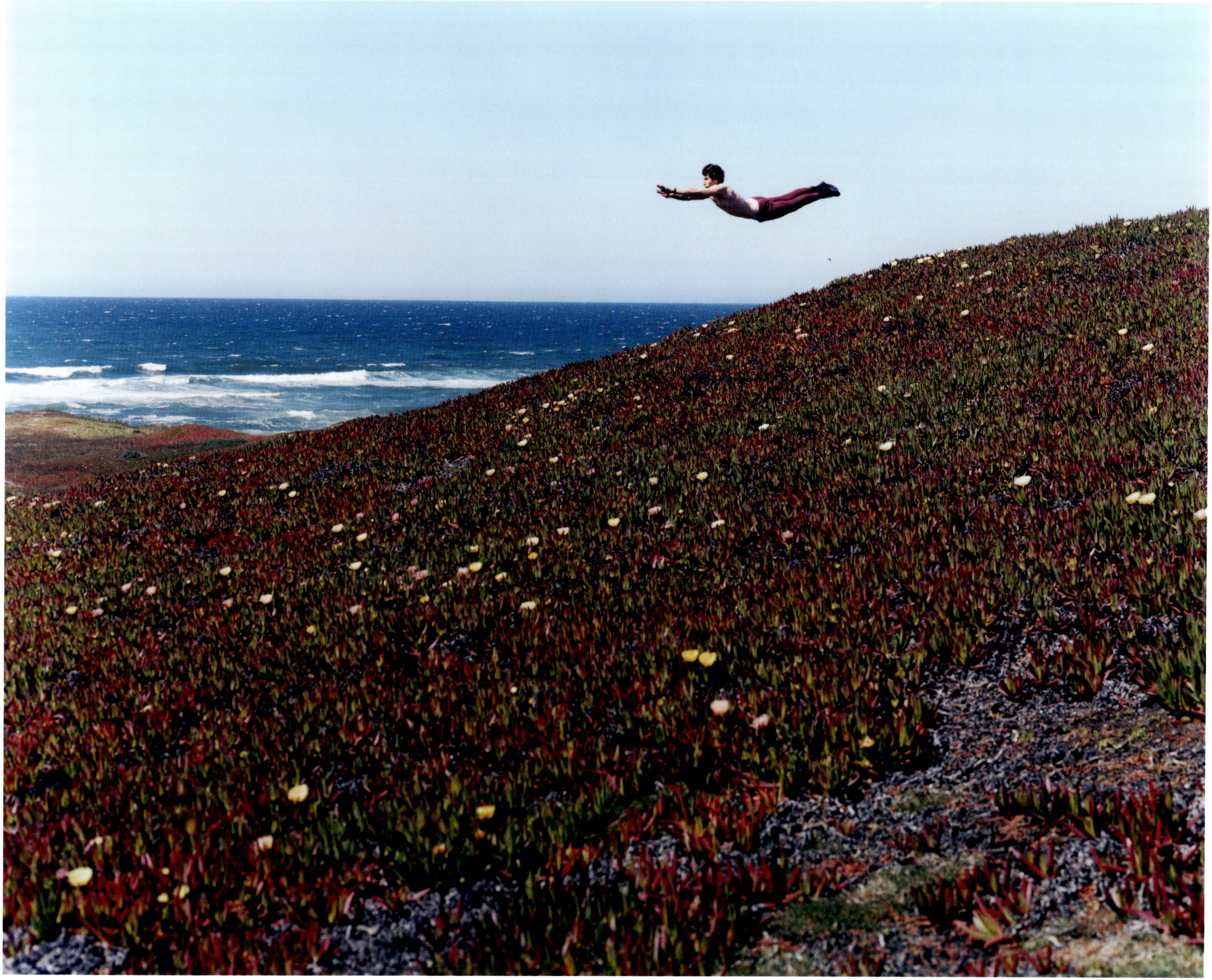

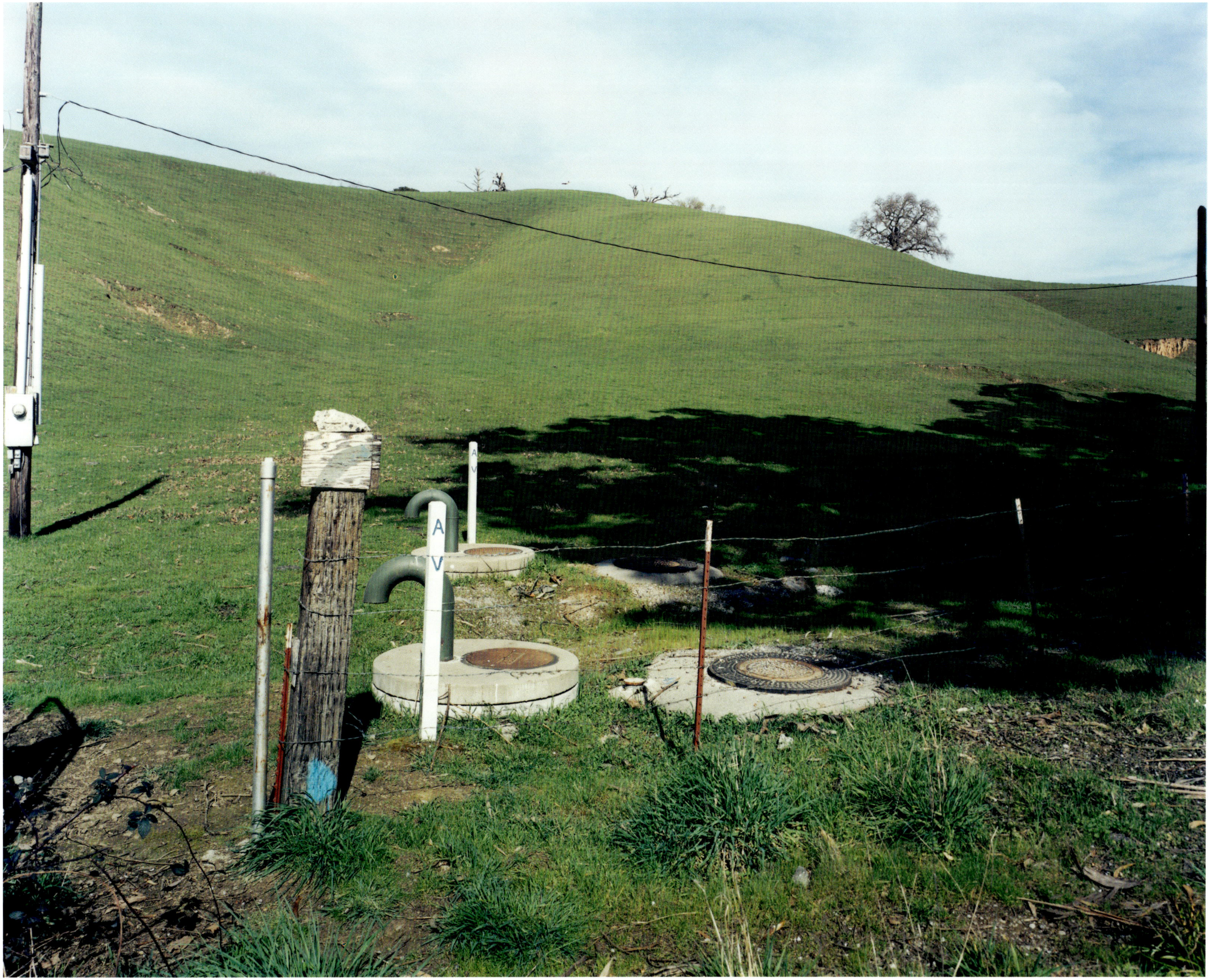
A
V

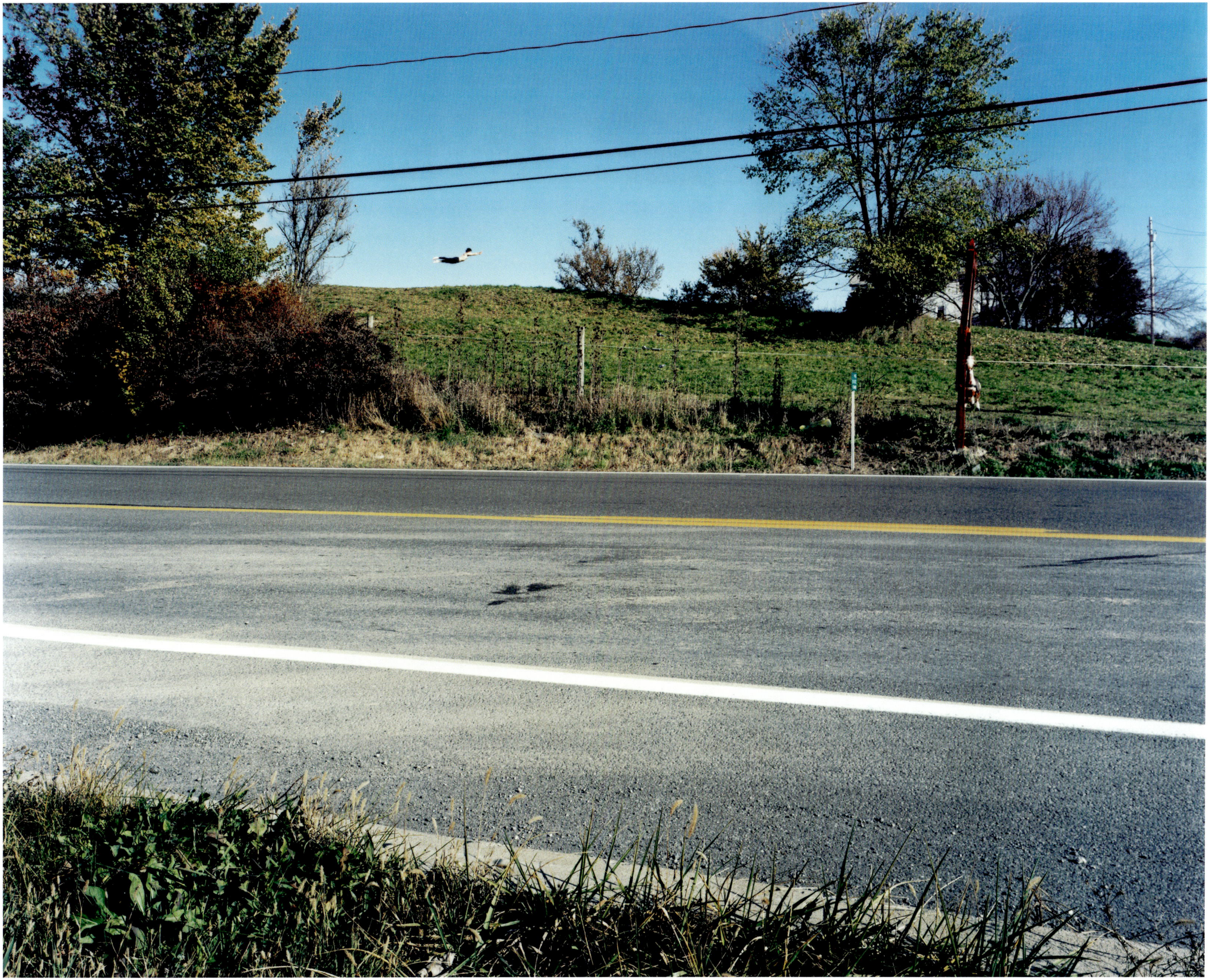

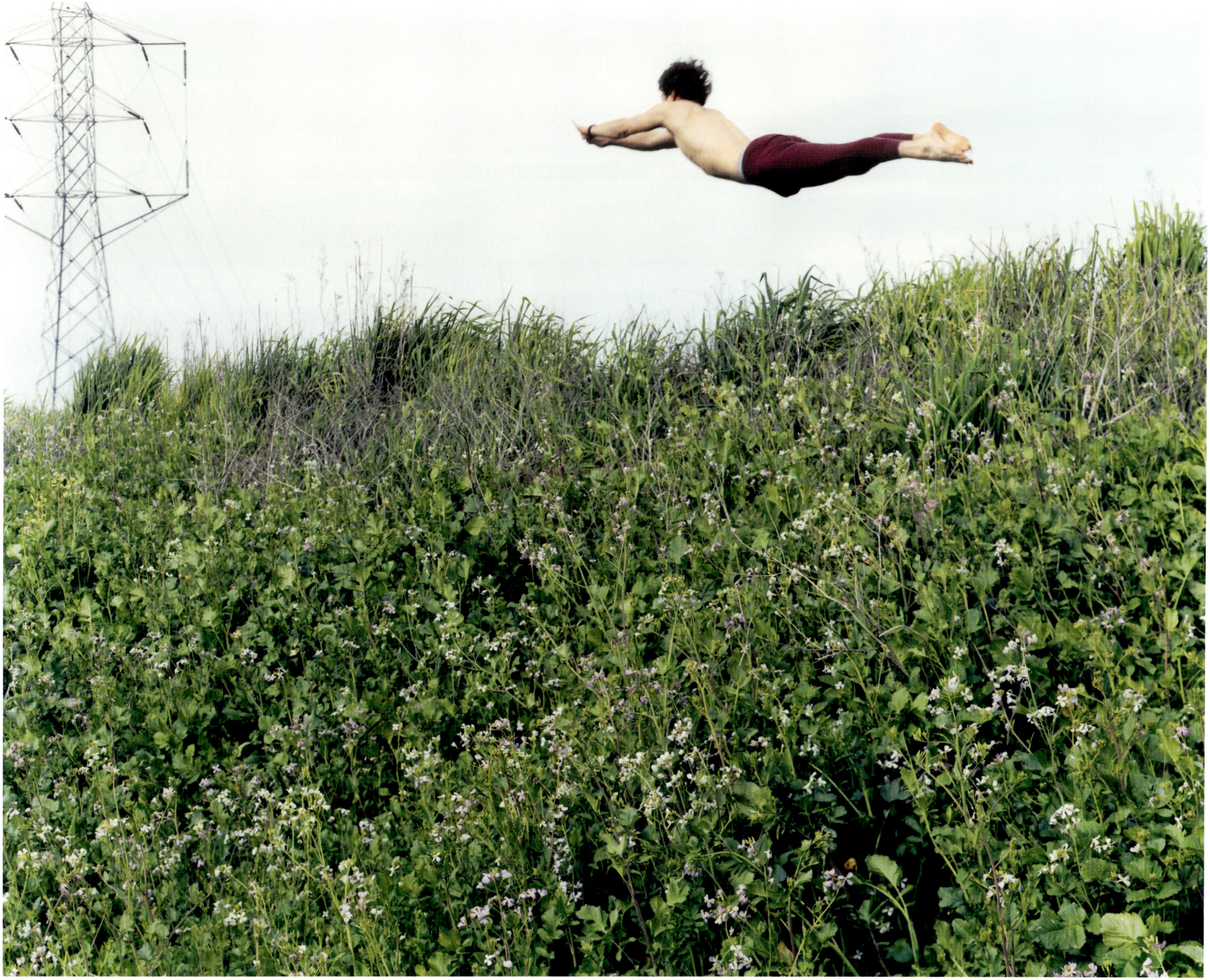

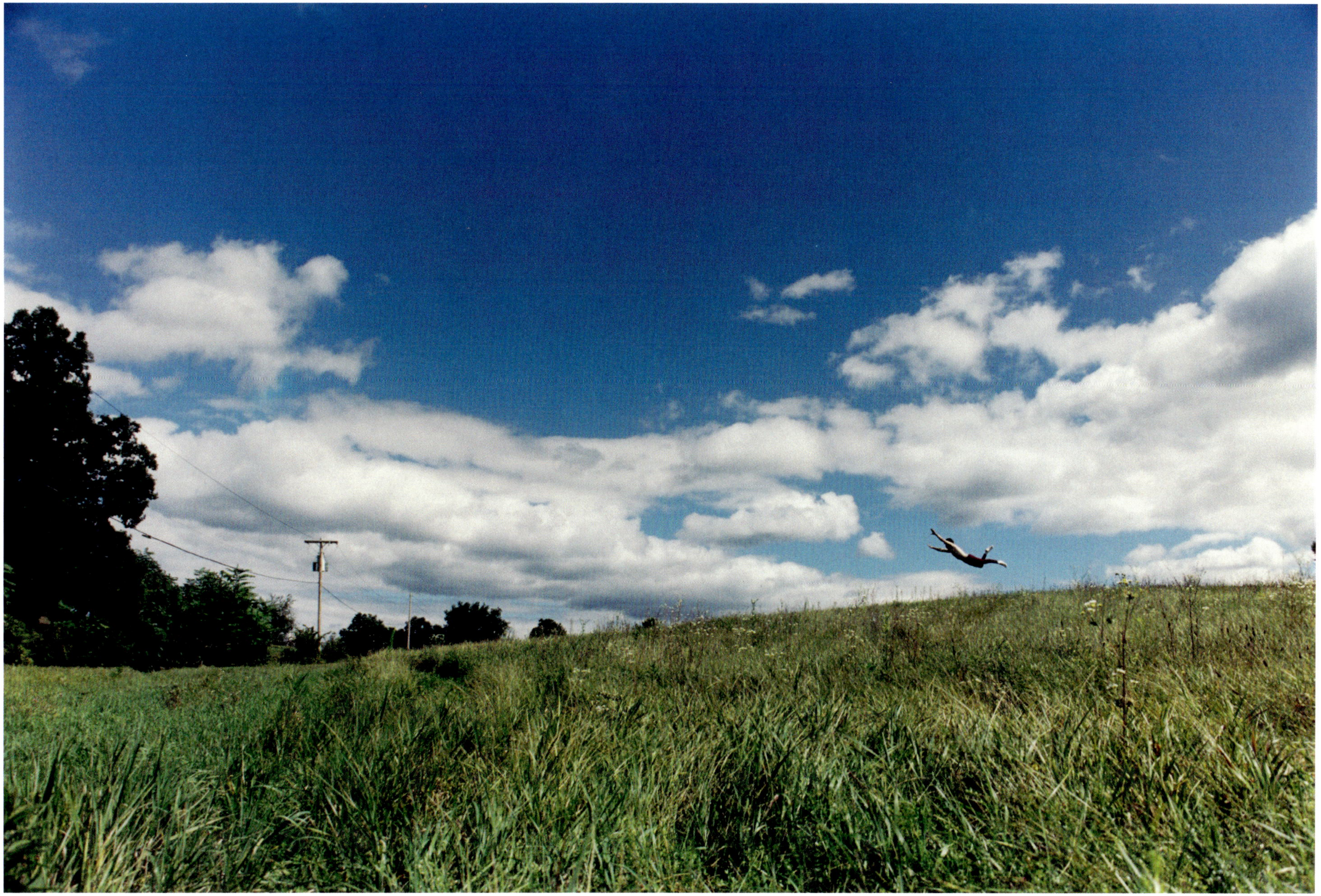

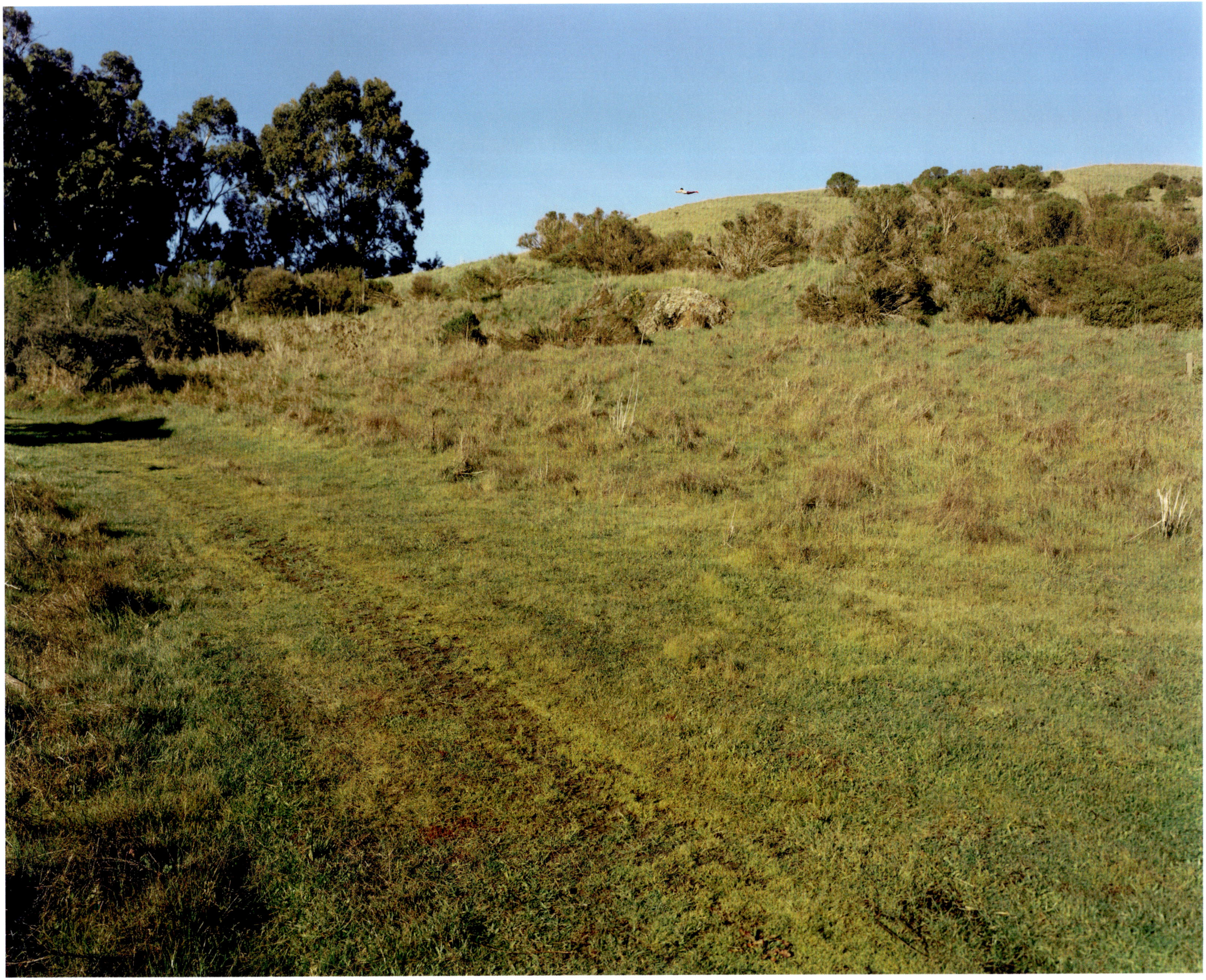

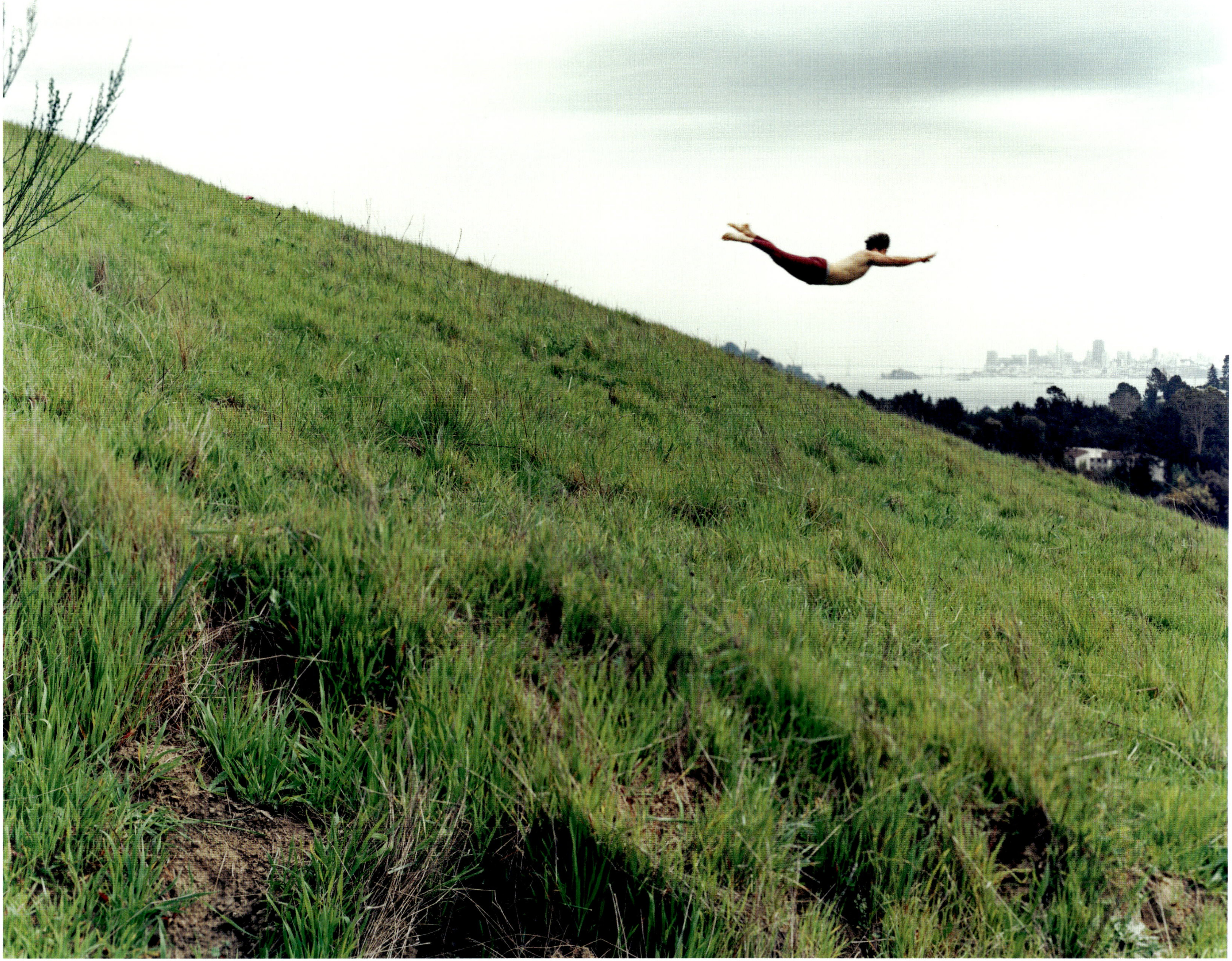

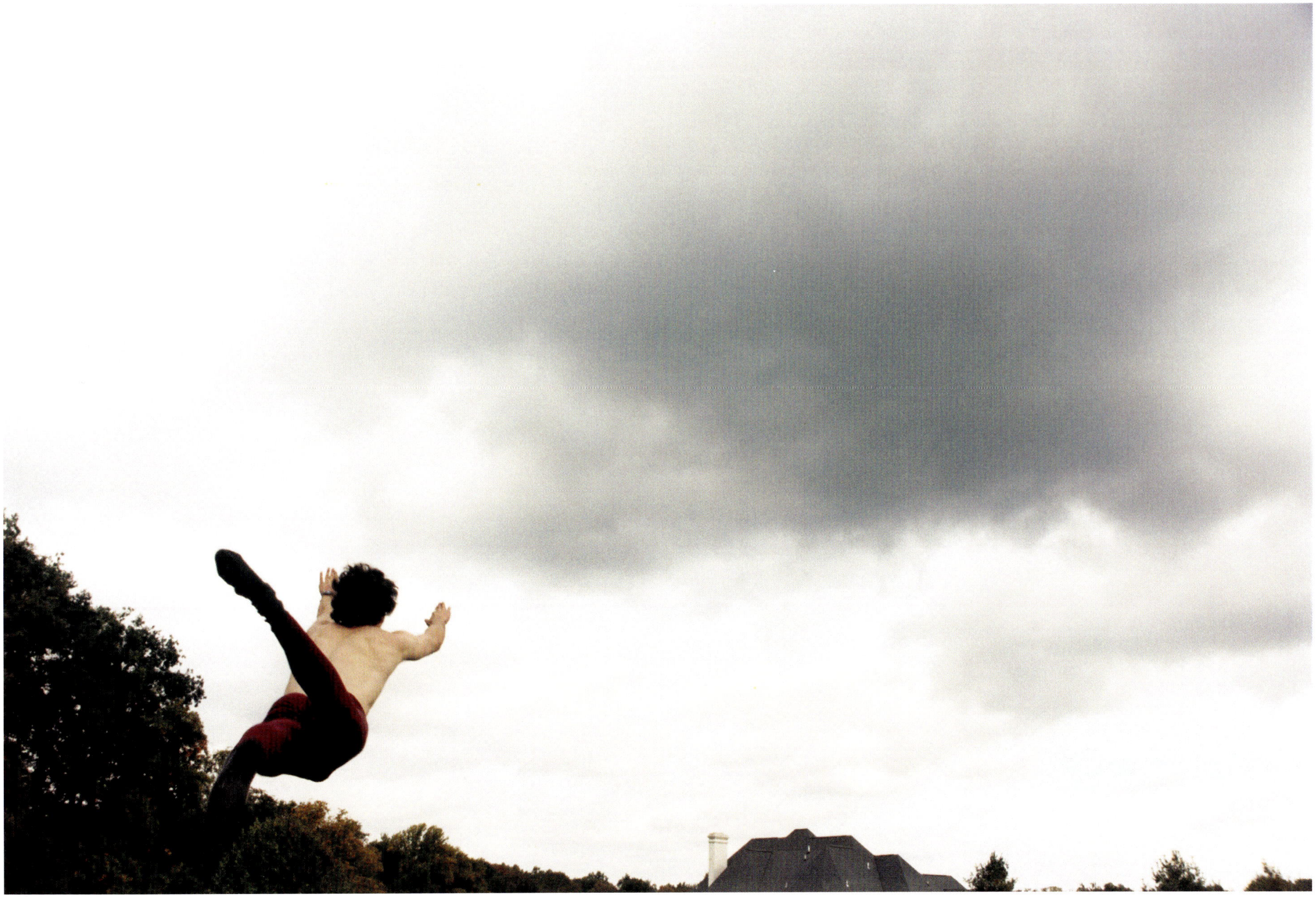

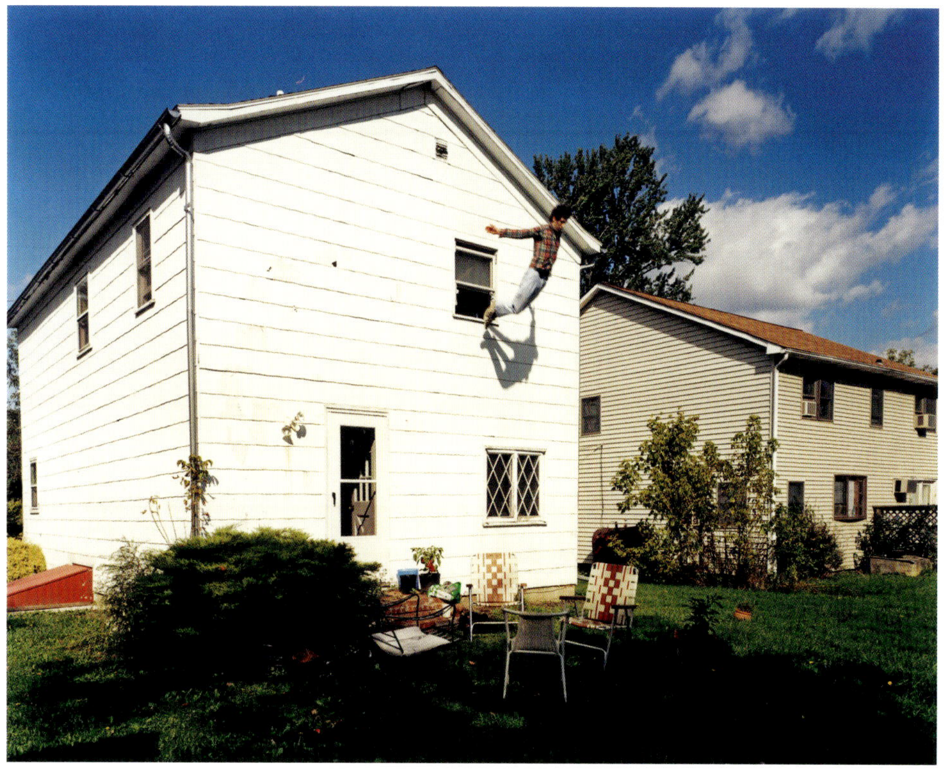

Acknowledgements

This book was made possible through generous support from the following people:

Stephen Shore and the Bard College Photography Department, John Pilson, An-My Lê, Larry Fink, Barbara Ess, Laurie Dahlberg, Len and Chandra Gordon, Nirah Shirazipour, Erin Miksic, Sarah Wymer, Ted from Oklahoma, Ruby Sky Stiler, Gary Kornblau, Kate Wolf, Kolby Yarnell, David Quadrini, Kelly Taxter, Pascal Spengemann, Grace Evans, Zach Feuer, Leo Koenig, Lou Reese IV, Mei Mei Hu and the Reese Family, Claudia Groeflin, Davia Maag, Iwan Wirth, Craig Cohen, Kiki Bauer, and everyone at powerHouse Books.

I would also like to thank Gregory Crewdson for his generous contribution and Cosi Theodoli-Braschi for her help in writing his foreword. And lastly, I'd like to thank my grandfather, Don Geiger, for his words.

Flying Pictures

Published in the United States by powerHouse Books,
a division of powerHouse Cultural Entertainment, Inc.
37 Main Street, Brooklyn, NY 11201-1021
telephone 212 604 9074, fax 212 366 5247
e-mail: flyingpictures@powerHouseBooks.com
website: www.powerHouseBooks.com

First edition, 2009

Flying Pictures has been printed in a limited edition of 1,000 copies.

Library of Congress Control number: 2009929084

Hardcover ISBN 978-1-57687-513-1

Color separations by FCI Digital

Printing and binding by Everbest Printing Company through Four Colour Imports, KY

Book design by Kiki Bauer

Epigraph from the poem "Speed of Sound" collected in *Unexpected Truce*
(1965, Round Table Books)

A complete catalog of powerHouse Books and Limited Editions is available upon request; please call, write, or visit our website.

10 9 8 7 6 5 4 3 2 1

Printed and bound in China